How to Choose Your Perfect Engineering Career

Cathleen Small

CHERITON CHILDREN'S BOOKS

Published in 2023 by **Cheriton Children's Books**
PO Box 7258, Bridgnorth, Shropshire, WV16 9ET, UK

First Edition

Author: Cathleen Small
Designer: Paul Myerscough
Editors: Sarah Eason and Jennifer Sanderson
Proofreader: Ella Short

Picture credits: Cover: Shutterstock/SpeedKingz. Inside: pp. 1, 11, 19: Shutterstock/SeventyFour; pp. 1, 45: Shutterstock/Monkey Business Images; p. 4: Shutterstock/Krakenimages.com; p. 5: Shutterstock/Fizkes; pp. 6, 10: Shutterstock/Zapp2Photo; pp. 6, 21, 28: Shutterstock/EfteskiStudio; pp. 6, 30, 38: Shutterstock/Goodluz; pp. 7, 40, 49: Shutterstock/Bixstock; pp. 7, 54, 59: Shutterstock/Hodoimg; p. 8: Shutterstock/SpeedKingz; p. 9: Shutterstock/Mangostar; p. 12: Shutterstock/Paskee; pp. 13, 18: Shutterstock/Chaiviewfinder; pp. 14, 18: Shutterstock/Monkey Business Images; p. 15: Shutterstock/Viktoriya; pp. 16, 19: Shutterstock/Subin Pumsom; p. 17: Shutterstock/Chaay Tee; pp. 20, 28: Shutterstock/Angellodeco; p. 22: Shutterstock/Doug Lemke; pp. 23, 29: Shutterstock/Monkey Business Images; p. 24: Shutterstock/PopTika; pp. 25, 29: Shutterstock/SpeedKingz; p. 26: Shutterstock/TuiPhotoEngineer; p. 27: Shutterstock/M G White; p. 31: Shutterstock/Nejron Photo; pp. 32, 39: Shutterstock/Gorodenkoff; pp. 33, 39: Shutterstock/NDAB Creativity; p. 34: Shutterstock/RedPixel.PL; p. 35: Shutterstock/Damir Khabirov; p. 36: Shutterstock/LightField Studios; pp. 37, 38: Shutterstock/Monkey Business Images; pp. 41, 48: Shutterstock/Fizkes; p. 42: Shutterstock/Ververidis Vasilis; pp. 43, 48: Shutterstock/NDAB Creativity; p. 44: Shutterstock/Gorodenkoff; p. 46: Shutterstock/Fizkes; pp. 47, 49: Shutterstock/Fizkes; p. 50: Shutterstock/Chaosamran Studio; pp. 51, 58: Shutterstock/Gorodenkoff; pp. 52, 58: Shutterstock/Kzenon; p. 53: Shutterstock/SatawatK; p. 55: Shutterstock/Gorodenkoff; p. 56: Shutterstock/NDAB Creativity; pp. 57, 59: Shutterstock/NDAB Creativity; p. 60: Shutterstock/Fizkes; p. 61: Shutterstock/Rawpixel.com..

Printed in China

Please visit our website,
www.cheritonchildrensbooks.com
to see more of our high-quality books.

CONTENTS

CHAPTER ONE
WHO DO YOU THINK YOU ARE? 4

CHAPTER TWO
HELPER ROLES IN ENGINEERING 10

CHAPTER THREE
BUILDER ROLES IN ENGINEERING 20

CHAPTER FOUR
CREATOR ROLES IN ENGINEERING 30

CHAPTER FIVE
ORGANIZER ROLES IN ENGINEERING 40

CHAPTER SIX
THINKER ROLES IN ENGINEERING 50

WHAT NEXT?–YOUR CAREER CHECKLIST 60

GLOSSARY 62
FIND OUT MORE 63
INDEX AND ABOUT THE AUTHOR 64

WHO DO YOU THINK YOU ARE?

If you are a student, there is probably a lot you still do not know about yourself—even when you are an adult, you will still have things to learn about yourself. But you may know that you are interested in an engineering career, and that is why you are reading this book.

Finding Your Path

Engineering is a constantly growing field with many different exciting career paths that you can follow. Do you have an idea of what you want to do in the engineering world? You might but if not, don't worry—there is plenty of time to figure out what you want to do. This book will help you take the first step in deciding what to do next. It will also help you start to define your interests in the larger field of engineering and narrow down your career options.

Today, engineering careers feature in almost every part of our lives, from the buildings we live and work in to the transportation systems we use to get to and from home and work. With so many choices, how do you start narrowing down your search for a career in engineering? One way to begin is by identifying your personality type and then looking at the types of careers that are often a good fit for people who share the same personality type as you. So let's get started!

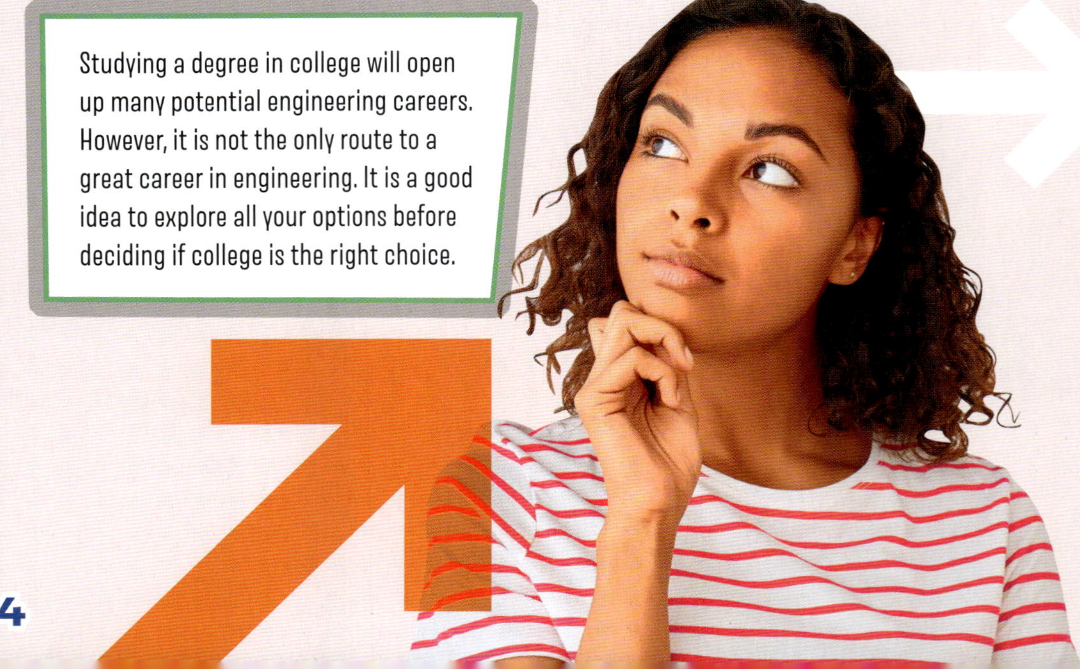

Studying a degree in college will open up many potential engineering careers. However, it is not the only route to a great career in engineering. It is a good idea to explore all your options before deciding if college is the right choice.

Follow the Flowchart to Fast-Track Your Future!

To find your personality type, you'll need to work through the flowchart on the following pages. The flowchart asks you simple questions to help you determine your personality type. This is based on five types of personality and the work situations that typically appeal to people with those types of personality. The five personality types in the flowchart are: social; practical; creative; organized; and analytical.

These personality types correspond to five different work environments. Social people, for example, often enjoy helper types of careers. Practical people often thrive in builder types of jobs. Creative people flourish in creator roles. Organized people typically enjoy organizational roles. And analytical people usually gravitate to thinking jobs. Although flowcharts are not foolproof, the one in this book should help steer you in the right direction.

Exploring your personality type and career options is important because if you work in an area that is a good fit for you, you will be more likely to be satisfied and successful. And that, in turn, will lead to greater overall happiness with your life. So take a look at the flowchart on the next pages and work through it to determine your career personality type. Then, follow the next steps to find your perfect career in engineering.

This flowchart asks you questions about your preferences to help you figure out which of the five personality types best describes you. It helps you think about what you like and don't like, and what kind of work might be best for you, so you can make sound career choices.

Once you have figured out what your career type is, take a look at the career choices in this book. Each chapter features some interesting careers linked to the personality types shown in the flowchart on these pages. A variety of jobs are explored in each chapter, along with a day in the life of one of the roles featured. Each chapter concludes with a checklist that helps you work through how you feel about the featured jobs and if they may be right for you.

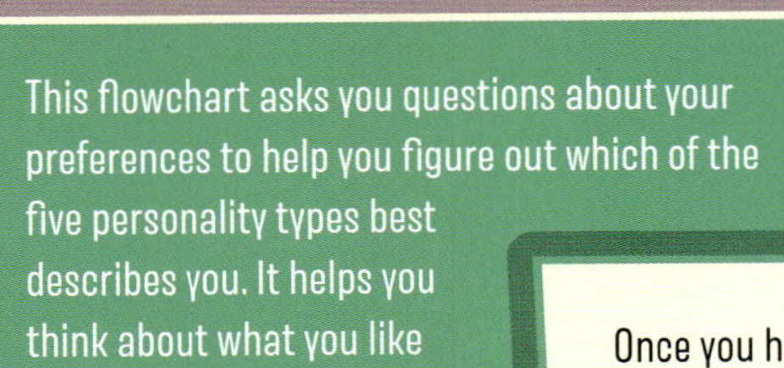

Are you interested in helping people and the environment?

No, this is not an area I want to work in

Yes, I want to make a positive contribution to people's lives and the environment

Do you like working in a practical, hands-on way?

No, I hate having to be practical

Yup, I am super practical

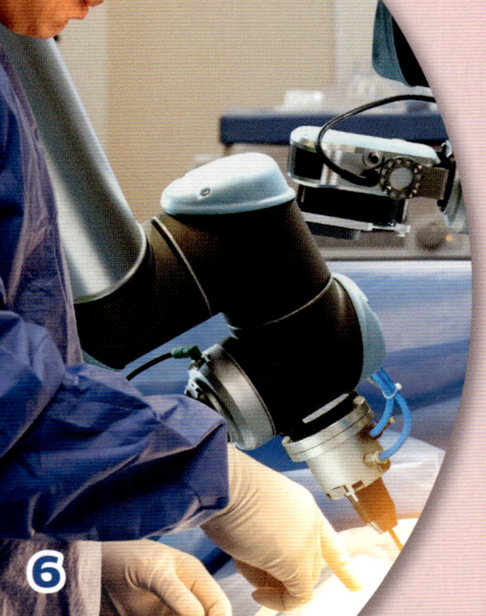

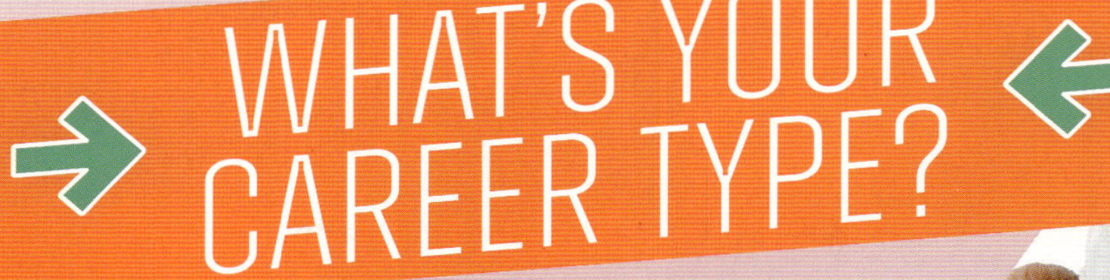

Creator

Organizer

Thinker

Yes! Check that box!

Do you love having things in order?

No, art's not my thing!

That's me!

No, I don't care about that

Are you artistic or creative?

Do you enjoy studying and thinking through complex ideas?

No, I don't enjoy complicated study

Yes, I love theories and thinking through ideas

If you have made it all the way to this box, try taking the test again— you can work through it a few times before you finalize your answer.

What Did You Learn?

Did you discover anything new about yourself from the flowchart? Did you find a career personality type that suits you? If you found a perfect fit, that's great but if you didn't, don't worry—most people find they are a combination of more than one type of personality. For example, creative people can be helpers, too, and organized people are often thinkers. You do not have to be just one type.

Weed Out the Nos

If, after following the flowchart, you still feel a little unsure about your career personality type, give it another try but this time, don't worry about fitting into just one career personality type. Rather, weed out the definite nos, and then look at the remaining possibilities. Think about which ones might be the closest fits.

What to Do with the Yesses

Once you have narrowed down your career personality type or types, if you're a combination of a few, work through this book to look at the engineering careers that are well suited to each personality type. Each chapter is devoted to specific engineering fields that typically work well for a given personality type. Start with the chapters that deal with the career personality types that best fit you but be sure not to miss the other chapters, too. You never know when a career in a different personality group might resonate with you, so it is a good idea to stay flexible and open to ideas.

Explore Options

This book covers many options in the field of engineering but not all of them. You can read the book to get started and then explore further. The Review and Check In sections at the end of each chapter feature further career options you could research. And once you reach the end of the book, the What Next? checklist will guide you through taking the next steps to kick-start your career.

Keeping an open mind about your future choices will give you space to explore a lot of different options, and find the ones that best suit you.

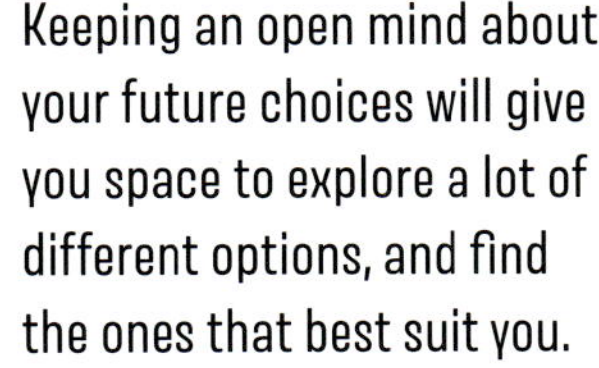

CHAPTER TWO

HELPER ROLES IN ENGINEERING

Maybe you would like to help a specific group of people, or perhaps you enjoy helping people in a teaching or mentoring role, or maybe you want to do something that makes a difference to a collective group of people, such as saving the environment for future generations. Either way, if you scored highly on the "helper" career personality type, this chapter has a lot of useful information for you. Let's take a look.

Engineering Problem Solver

There are many ways to help people. Some people shape their career around helping specific groups of people—for example, those who want to make a difference to people who are homeless may work at job-training programs for homeless individuals or work at shelters for homeless people. But, how can you combine your desire to help such people with engineering?

Start by thinking about objects or machines that have helped people. For example, toilets and indoor plumbing have made people's lives a lot easier. Hundreds of years ago, before toilets and indoor plumbing were invented, people used chamber pots instead of toilets. They would empty them out each day, usually just by dumping them out in a field or on the street. It was not a very sanitary

Smart engineering solutions help surgeons perform delicate operations with accuracy.

If you become an engineering problem solver, you could help develop prosthetic limbs that better meet patients' needs.

process and as a result, illnesses, such as cholera and dysentery, were spread. The toilet may seem part of everyday life but it was an engineering solution that helped save lives. Even now, people in some rural areas and developing nations do not have access to modern plumbing systems and toilets, and disease is still very much a concern because of that.

You could use your engineering skills by looking at people who are somehow lacking in a basic human need and trying to figure out a solution that could be engineered to solve the problem. For example, you might be interested in engineering solutions to improve the living conditions of people who live in extreme poverty. Or, maybe you want to

think globally, and engineer solutions to help people in developing nations. Or you can think about medical needs and how solutions for them could be engineered. The medical field is constantly evolving, with new machines and devices designed to help patients being developed all the time. Prosthetic limbs are a good example, as well as devices such as mechanical heart valves and robotic machines, which assist surgeons with delicate surgeries.

Most solutions that help people and contribute to the greater good have some level of engineering behind them. If this type of career appeals to you, there are many exciting opportunities that you can explore.

Career Insight:
Shelters for Refugees

In 2015, two graduate architecture students at Rice University in Houston, Texas, turned their desire to help others into a nonprofit company called Every Shelter, which has resulted in more than 480 refugee families obtaining shelters with adequate flooring. Shelters in refugee camps in the Middle East often have no floor, which means refugees are sleeping on dirt, which is both cold and unhygienic. The pair of students who started Every Shelter began their careers working for large architectural firms on projects local to the Houston area, working on their flooring project on the side but as their small project grew bigger, they were able to focus their efforts on their project with refugees instead. Today, they are continuing to provide an engineering solution that has created a positive impact on the lives of thousands of refugees.

Refugee shelters are usually very basic. Camps that can provide adequate housing help reduce the hardship experienced by refugees.

Environment Engineer

Another way to find a helper role in engineering is to work in engineering fields that benefit the environment. We all live in the environment, and one day, your children and grandchildren will too. For that reason, it is important to preserve the environment as much as possible and try to undo some of the damage that may have been caused by past generations. For example, alternative energy solutions are in high demand because for hundreds of years, fossil fuels such as coal were used as an energy source, having long-lasting effects on the environment.

Within the past couple of generations, people have realized that if we want life on our planet to continue for future generations, we need to find other energy solutions that are renewable and cause less harm to the environment. This has opened up the vast field of renewable energy engineering. Engineers in this area work on alternative energy sources such as wind power, solar power, hydropower, geothermal energy, and biomass. You may have seen the wind turbines used to harness the energy of wind. Engineers designed those turbines and are constantly refining them for maximum efficiency. The same is true with solar power—engineers work to develop ever more efficient solar panels and solar cells to harness the energy from the sun and turn it into usable electricity. Engineers also design the hydroelectric plants where the power of moving water is harnessed to generate electricity. And geothermal energy comes from beneath Earth's surface and requires engineered heat-pump systems to harness it and convert it into usable power. Engineers continue to design solutions to convert biomass into electricity or biofuels that automobiles and other engines can run on.

To become a renewable energy engineer, you'll need to be passionate about the environment and sustainable living. You will also need a bachelor's degree in mechanical, electrical, chemical, or environmental engineering. You may also need certification from a state licensing board.

Another way to help people through a career in engineering is to teach. You can teach engineering at college or university level, assuming you have achieved the proper level of education yourself. Most colleges and universities require teachers to have at least a master's degree in their field, although many require a PhD.

Alternatively, if you have a bachelor's degree in engineering but feel passionate about teaching future engineers, you could consider teaching engineering classes at high school level. Not every high school will have engineering classes but many will have engineering clubs or extracurricular programs that are staffed by teachers with expertise in engineering. There are also

If you enjoy communicating ideas, you could combine that with your passion for engineering by teaching.

Career Insight:
Innovative Engineering

Certain fields of engineering, including mechanical, structural, and chemical engineering, have been around for a very long time. However, there are some new and innovative engineering programs out there, too, for people looking for something a little bit different. For example, Carnegie Mellon University (CMU) in Pittsburgh, Pennsylvania, offers an engineering program in robotics. Louisiana State University in Baton Rouge offers a disaster science and management program. Tufts University on the border of Medford and Somerville, Massachusetts, offers a degree in engineering psychology, which looks at the science of human behavior and ability, and applies it to technology. In Denver, Colorado, students who are interested in designing buildings to be as energy efficient as possible can work toward a green interior design qualification at the Rocky Mountain College of Art + Design.

summer engineering camps for students, again staffed by people with experience in the engineering field. And there are community classes and programs for children with an interest in engineering.

What you will be qualified to teach depends on the level of engineering experience you have, what degree you have earned in engineering, and where you want to teach. Most public high schools, for example, require teachers to have a teaching credential, so you might have to go back to school for that. However, the credential program is not overly long, so if you decide teaching is what you want to do, obtaining a credential after you have already done your engineering coursework is not that much more work.

One thing to keep in mind is that teaching careers typically are not the highest-paying option. An engineer working in the field will most likely make more money than an engineer who is teaching engineering. However, teaching has benefits: many people enjoy the interaction with students —especially those who enjoy talking to people, since engineering can be rather solitary work—and you will also have long summer vacations, just like your students.

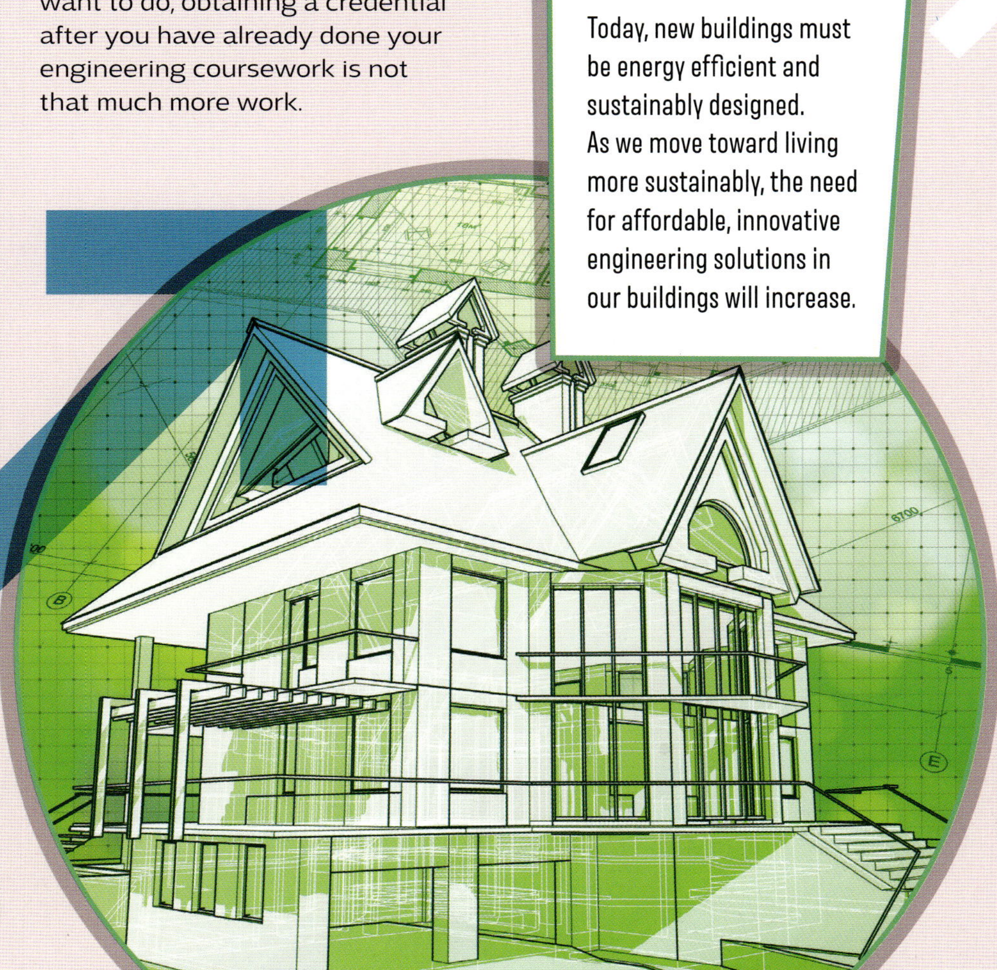

Today, new buildings must be energy efficient and sustainably designed. As we move toward living more sustainably, the need for affordable, innovative engineering solutions in our buildings will increase.

A DAY IN THE LIFE OF

A Structural Engineer

There are so many job roles in engineering but what is it like to spend a day as an engineer who helps people? Let's take a look at a day in the life of a structural engineer to find out more.

7:00 a.m. You like to get to the office bright and early to get organized for the day. The first thing you do when you arrive is review your calendar and check your emails and messages.

8:00 a.m. Once a week, you meet with your coworkers to talk about new projects coming down the pipeline. Today was a good meeting because you are named as the lead engineer on an exciting new project that will help an overcrowded local homeless shelter.

9:00 a.m. You leave your office for a site visit across town at a new shelter, which is being constructed to house more of the city's homeless population. You like site visits because they're a chance to see how things are taking shape and to engage with the people working on site.

9:30 a.m. You meet with the construction crew on site. Everyone is present, so you can review the crew's progress and troubleshoot problems that have come up. There's a problem with one of the steel suppliers, so once you're back in the office, you're going to have to talk to some suppliers to sort out the issue.

It is exciting to visit the construction site to see the build in development.

11:30 a.m. The emails that came in while you were on site are going to have to wait because you need to spend time on the phone with the steel supplier. The construction beams that were delivered have a design flaw that has made them unusable for the project. The steel supplier cannot fix the issue for at least four weeks, which will put the project far behind schedule, so you need to call other suppliers to find one with the appropriate beams.

1:00 p.m. No local supplier has the correct beams, so you spend time calling out-of-state vendors to find them. Finally, you locate one with the correct beams and spend time arranging transportation to the job site. You grab some food and come back to your desk, answering emails while you eat your lunch.

1:30 p.m. With the problem of the steel sorted, you feel relieved and start working on plans for the exciting new project you are leading. The head of the firm wants to see a preliminary design by Friday so you have a lot of work ahead of you.

6:00 p.m. You arrive home for dinner with your family. You may do some work later because your new project is really exciting. It is not an easy job but you are comfortable in the knowledge that you are playing a big part in helping the homeless in your city.

Helper

Review and Check In

Now that you've learned about some of the helper roles in engineering, explore more about how you feel about each job. Use the questions and checklist below to help you assess how strongly you feel each role might suit you.

Environment Engineer

- Why do you want to make a difference by working as an environment engineer?
- What field of environmental engineering interests you most? Why do you say so?
- Are you an innovative problem solver? What other skills do you have that would make you a good candidate for doing this job?
- Is there anything that you do not like about the role? What is that? Do the benefits of the job outweigh the aspects you may not like?
- Environment engineers often work in the field to check on their work. Why would this suit you?

Engineer Educator

- What about being an educator would you find rewarding?
- Is there a particular age group or level of education you would prefer to teach?
- Are you confident when speaking in front of others? If you aren't, how could you improve on this? What other challenges do you think you'll face doing this job?
- Would you enjoy studying further to make sure you are always knowledgeable in the latest engineering discoveries? Why do you say so?
- Would you be happy to earn a lower salary relative to other areas in the field in order to have a job that educates people?

"Always keep in mind what you want from your career."

A Structural Engineer

- Review the Day in the Life feature. Did anything about the day surprise you? What was that?
- What do you think the challenges were, and how might you deal with them?
- What aspects of the role do you think you would enjoy?
- Look at the structure of the day and the working hours. Would you be happy with that structure? How might it affect your lifestyle?

Research More

If you like the idea of working as a helper in engineering, but are not sure the jobs covered in this chapter suit you, here are some more helper roles you could explore.

Customer Service Engineer
Engineering Apprentice Development Coach
Forestry Mechanical Engineer

CHAPTER THREE

BUILDER ROLES IN ENGINEERING

Builders typically like to work in a hands-on way. Sometimes, that means using their hands to build, like people who work in construction, other times that may entail building via machine or computer, such as people who create computer-generated models for structures. If you are a person who likes hands-on work, there are plenty of engineering opportunities available for you.

Biomedical Engineer

The term "biomedical" usually brings to mind biology and working in a lab but biomedical engineers also build or maintain medical devices such as X-ray machines, vision-testing devices, prosthetics, and many other sorts of medical gadget. People working in biomedical instrumentation develop instruments and therapeutic equipment. Medical equipment requires servicing and repair just like any other mechanical device: your computer, a car, and so on—and biomedical service engineers are the people who service and repair this equipment.

Some biomedical engineers also work on pharmaceutical drugs, which is perhaps less of a traditional "builder" role but still involves creating something in the field. However, there are many opportunities in biomedical engineering for those who want to be really hands-on, working in hospitals, universities, manufacturing plants, government regulatory agencies, and research facilities.

To work as a biomedical engineer, you typically need a bachelor's degree in bioengineering, biomedical engineering, or in another related engineering field. Some positions also require a graduate degree.

Biomedical engineering roles can be diverse, from helping create medical equipment through to developing new drug therapies.

Civil Engineer

Civil engineers are often involved with the design and inspection part of constructing structures and systems, such as roads, airports, tunnels, dams, and buildings. However, they do work hands-on sometimes, depending on the project. Becoming a civil engineer requires a bachelor's degree. For those interested in eventually moving to a senior or management position, a master's degree is usually required, along with licensing via the Professional Engineer (PE) exam.

If you enjoy the idea of working on structures but you are not sure you want to earn a bachelor's degree, you might want to become a civil engineering technician. This position allows you to assist a civil engineer in planning, designing, and building structures and systems.

It requires only an associate's degree, which is typically a two-year degree compared with the four-plus years needed to complete a bachelor's degree. Engineering technology programs at vocational schools can also provide a sufficient education for a civil engineering technician.

Just keep in mind that if you want to eventually move into a senior or management engineering position, which will also bring a greater salary, you will need to invest in your education and earn bachelor's and master's degrees, along with the appropriate license for your state.

The job outlook is good for both civil engineers and civil engineering technicians—both career areas are expected to grow slightly faster than the average projected growth across all occupations.

Career Insight:
Adrenaline Engineer!

You might think of civil engineers as primarily working on buildings, bridges, and similar structures but there are some rather unique opportunities available for civil engineers, too. For example, those who thrive on adrenaline might want to consider being a mechanical engineer who specializes in roller-coasters.

Roller-coasters are exciting for many—and the faster and scarier the better! Roller-coaster engineers are the people who determine just how fast and how exciting those coasters can be. They weigh up safety concerns, the environment where the coaster will be built, and the design of the track when designing and building the most appealing coasters for thrill-seeking customers. Roller-coaster engineers do a lot of testing, too, because they must ensure that the ride is safe enough for riders. Coasters need to accelerate fast enough to climb hills but not so fast that riders will be in danger of getting injured. The curves must be just the right angle to keep the coaster on the track but engineers also have to make sure the turns are not so tight or abrupt that riders get whiplash. It is a challenging job but the end result can be a ride that thrills thousands of people.

Being a "builder" engineer can involve working on some amazing projects —including theme park rides.

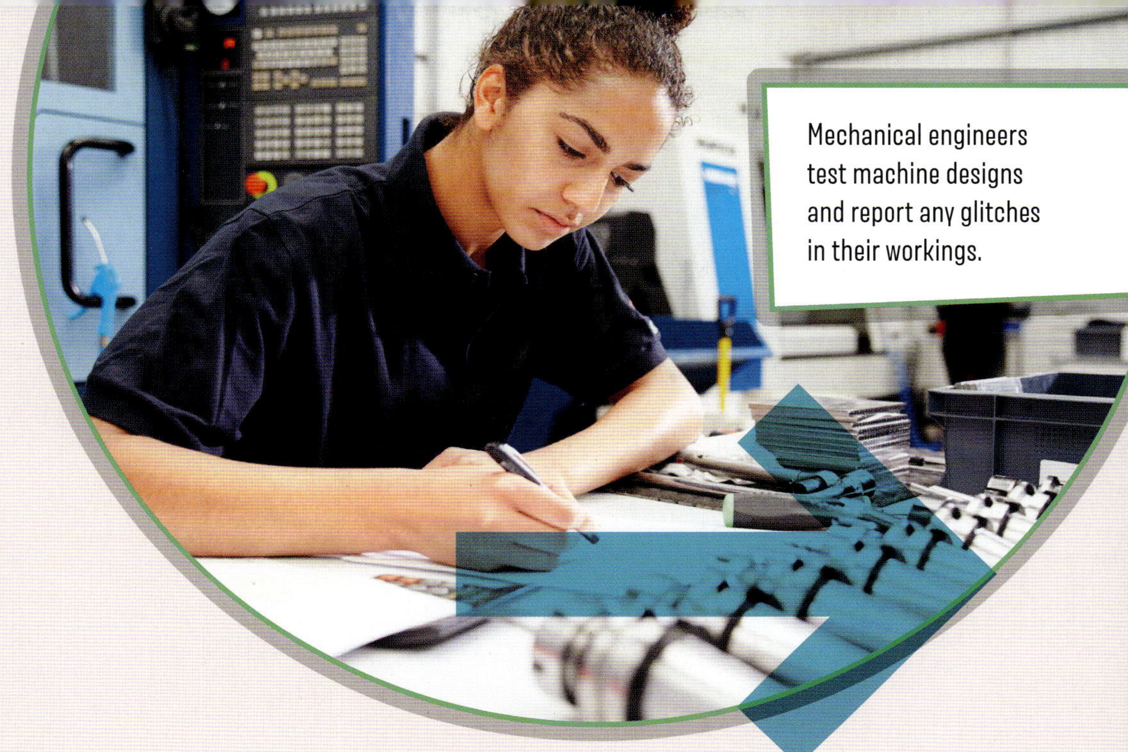

Mechanical Engineer

If you are really interested in building things, mechanical engineering may be a good option to pursue. Mechanical engineers design, build, and test engines, tools, and machines. They must earn at least a bachelor's degree, and there are degrees in mechanical engineering and mechanical engineering technology available at many universities. For those who eventually want a senior or management position in mechanical engineering, a master's degree will be required.

There are also mechanical engineering technicians who assist mechanical engineers on designing, building, and testing machines. If you do not wish to pursue a bachelor's or master's degree right away, an associate's degree is all that is needed to work as a mechanical engineering technician. An associate's degree can be earned through community college or vocational programs.

Another related job path would be to become a machinist. Machinists are the people who operate the machines that produce metal parts, instruments, and tools. Machinists do not necessarily design things and do not build engines but they do build instruments and tools, so it is a good job for someone interested in hands-on building work.

Machinists also do not need a college degree, although some choose to attend vocational school or earn an associate's degree. But typically, all that is needed is a high school diploma and on-the-job training or an apprenticeship.

Monitoring and controlling robot machinists will become a more common engineering role in the future.

Machinist: What Does the Future Hold?

The one drawback to becoming a machinist is that job growth in the field is not nearly as fast as in certain other engineering fields. This is because a lot of machinist work is now done by computers or robots, which cuts down on the number of human workers needed to complete tasks. However, there are still machinist jobs that require humans, and workers do retire or leave the profession every year, so there will be opportunities, depending on your chosen field.

If you like working on cars, you might consider a career as an automotive service technician. For those who enjoy building and getting their hands dirty, repairing cars can be a fun and rewarding career. Although many automotive problems are diagnosed by computers now, repairs are still often done by skilled technicians. Service technicians do not need a college degree, though some choose to earn an associate's degree. Usually, a certificate from a vocational or technical school is all that is required and on-the-job learning. Often service mechanics start with one brand of automobile and learn how to service that brand before moving to another. If you have a lot of experience and the drive to run your own business, you could open your own garage, servicing a wide range of vehicles.

Aerospace Engineer

If you like to build things and you're interested in aeromechanics, you might want to consider a job in aerospace engineering. Aerospace engineers design airplanes, spacecraft, and satellites, as well as missiles. Once their designs are complete, they usually build prototypes of them for testing.

Like most other engineers, aerospace engineers need at least a bachelor's degree, however, because the field is quite specialized, a master's degree is preferred. To work in the field of national defense (on space programs, missiles, and satellites, for example), aerospace engineers also often require security clearances. In most cases, you must be a citizen of the United States to receive these security clearances.

Aero engineering can involve working on a multitude of machines, from helicopters and airplanes to drones and satellites.

A DAY IN THE LIFE OF
A Civil Engineer

Watching your project come to life as a civil engineer is very rewarding. Although the job is mostly project-based and no two days are ever alike, these pages will give you some insight into a typical day's work.

7:00 a.m. You like to get an early start, so you're up early and at work by 8:00 a.m.

8:00 a.m. The first thing you do every morning is check your emails, messages, and calendar. You like to plan out your tasks, even if they may change a little as the day goes on.

9:00 a.m. You meet with your coworkers to discuss the company's latest building projects. There's a new bridge being constructed across a major river–it is a multi-year project, and you've been asked to head it up.

10:00 a.m. You leave the office to drive to the site of the new bridge to get a feel for it. You'll need to come back with a team to do a thorough assessment and report on the project but it's good to get a first look at what you'll be working on over the next few years.

12:00 p.m. While you're out, you pick up some lunch before heading back to the office.

During a site inspection, you make notes of all your findings. You'll use them to write up a report when you get back to the office.

Inspection work can be dirty, especially if you are examining pipework.

12:30 p.m. You spend the drive back to the office thinking about the new bridge—it's a really innovative project so you're very excited. As soon as you arrive back at the office, you start sketching out some ideas for the bridge.

1:00 p.m. You put your sketches aside to start drafting a Request for Proposals (RFP) for an Environmental Impact Report (EIR) on the site of the new bridge. These reports take a long time to prepare, so it's important that you open up an RFP soon. Environmental consulting firms will put together proposals to do the EIR, and after the close of the RFP you'll select one.

4:00 p.m. There are dozens of emails waiting so you start tackling those. You see an email about a problem on a road development project—you'll need to figure out a solution quickly to make sure the issue does not cost the company too much time and money. You make a note to prioritize this first thing tomorrow.

5:00 p.m. You head home but know that tomorrow is going to be another busy day as you have other projects to catch up on.

Builder

Now that you've learned about some of the builder roles in engineering, explore more about how you feel about each job. Use the questions below to help you assess how strongly you feel each role might suit you.

Biomedical Engineer

- What field of medicine are you interested in? How could you use your skills to make a difference in this field?
- What skills could you bring to this job?
- Are you happy to work as a team member, collaborating with others on projects?
- Would you prefer to work in a hospital, university, or manufacturing plant? Why?

Civil Engineer

- Would you enjoy working on-site, checking on projects, or would you prefer to work at a desk for most of your day? Why?
- If things don't go according to plan, how are you likely to react? How might this impact your job?
- What do you think the challenges of the job might be?
- Would you be prepared to study further to take the professional engineer (PE) qualification?

"When choosing a career, always consider the career outlook—does your chosen area have potential in the future?"

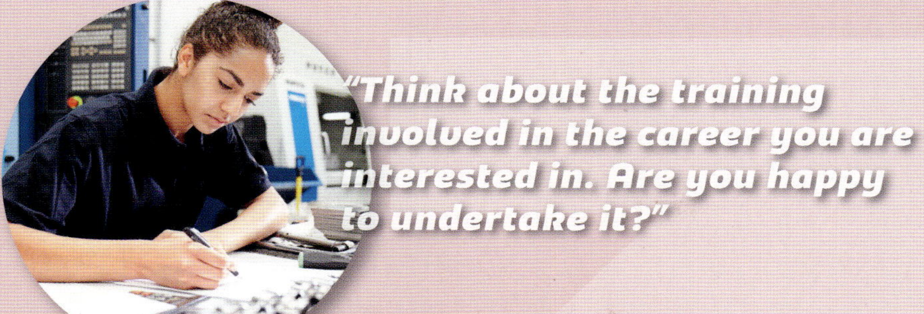

Mechanical Engineer

- What skills do you have that would make you well suited to this job?
- Often, mechanical engineers work long days and mostly in an office. Would this suit your personality? How might the long hours impact your lifestyle?

Aerospace Engineer

- This is a specialist role. Why would you be happy to study further to be qualified?
- Would you enjoy working for the federal government? Why?
- What particular skills do you have that you think could fit the role?

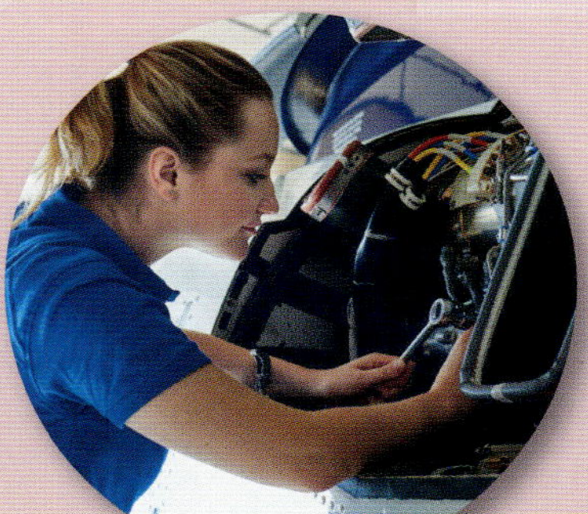

Research More

If you like the idea of working as a builder in engineering, but are not sure the jobs covered in this chapter suit you, here are some more builder roles you could explore.

Materials Engineer
Aerodynamicist
Avionics Hardware Engineer

CHAPTER FOUR

CREATOR ROLES IN ENGINEERING

Some might think there is little place for creative types in engineering, but that is not necessarily true. Those with artistic or creative talent can find some really interesting careers in the field of engineering. Let's take a look.

Industrial Designer

Industrial designers develop the ideas behind manufactured products. For example, think of the coffee machine or single-cup brewing systems. Years ago, they did not exist. Coffee makers just brewed whole pots of coffee. Then someone had the bright idea to design and build a coffee machine that would brew single cups of coffee, so you could have a hot, custom-brewed cup each time. Voilà—a new industry was born!

To become an industrial designer, you generally need a bachelor's degree in industrial design or some type of engineering. Depending on what you are designing, an architectural degree might work, too. Given how much work is done with technology today, you will need to gain experience with computer-aided design and drafting (CADD) software. Most universities will offer such courses but you can also take them at community colleges or vocational

Along with a flair for creativity, industrial design involves precise work and meticulous calculations.

Working as an audio engineer is a great way to combine your love of music and creativity with a passion for engineering.

programs. Experience with three-dimensional (3-D) modeling software is helpful, too. Again, most colleges will have classes you can take. There are also some free programs you can play around with, such as Blender. Many jobs do not necessarily require you to have learned such programs through school—it is fairly common for people to gain experience with software programs via hands-on learning and tutorials.

Audio Engineer

If music is your passion but you are also interested in technology, you might want to consider a career in audio engineering. Sometimes, these positions are called sound engineers or recording engineers but they all do generally the same thing: They create audio recordings of live performances by adjusting sound sources, mixing sounds, and implementing audio effects. If you have ever looked at a mixing board when the audio engineers are setting up for, or working at, a concert, you have undoubtedly noticed how many levers, knobs, and dials are on the board. The average person cannot just walk up to such a board, play with the knobs, and create a good or usable sound. It takes an engineer to know exactly where levels should be set and sounds should be balanced for the optimal listening experience. After all, no one likes to hear feedback when they are at a concert or if they are listening to a recording.

Audio engineers do not just record music or manage the sound at live concerts, though. They may also work on sound for television, film, radio, and video games, or for other live events such as theater productions and sporting events.

Become an Audio Engineer

While there are degree programs for audio engineers, a bachelor's degree is not always required. Many positions in this field only require an associate's degree. As with many fields, you can never go wrong by getting a degree but it is also not a hard-and-fast requirement if that is just not the right educational path for you to follow. For some, if you have an excellent ear for music, knowledge of how all the mixing equipment works, and can gain experience in the field through entry-level work, you can still have a future in audio engineering.

Acoustical Engineer

Acoustical engineers work on the actual space in which the sound is being made. For example, if you have ever attended a concert in a classical concert hall, you have likely noticed how good the sound quality is in every part of the room. That is because an acoustical engineer has designed the surfaces of the room to absorb sound where needed and allow the sound to fill the space where needed. Acoustical engineers also work to eliminate noise pollution in factories, construction sites, and businesses.

Acoustical engineers need a degree but as there are not many degree programs in acoustical engineering specifically, you can major in electrical engineering and then specialize.

Career Insight:
Full Sail University

If you are interested in a career that combines musical creativity with the science of engineering, Full Sail University may be of interest to you. The school, which is based in Florida, offers several degrees and qualifications including an audio arts undergraduate certificate and bachelor's degrees in audio production, music production, and recording arts. The campus has its own high-tech music venue, where students can learn to mix music on stage for live performances. Students also get to engineer recording sessions for professional artists and edit audio for games and shows. Full Sail also offers associate's, bachelor's, and master's degrees in many other related fields, such as film and television, digital animation, games, digital marketing, and software development.

Software Engineer

Software engineering is another good career path for creative people. Software comes in many varieties, from functional billing systems for hospitals, for example, through to highly creative apps that are used for leisure purposes, such as sport or entertainment. Sometimes, software engineers implement someone else's creative vision but other times, they come up with the vision. Today, apps have made it easier than ever to create software—now, almost anyone with a good idea and coding skills can build an app and market it.

In software engineering, people may work as part of a team to develop new programs. This growing field of work is great for creative minds that enjoy coding.

Do You Need a Degree?

The interesting thing about being a software engineer is that you don't necessarily have to have a college degree to do the job and be successful. Some companies will require a college degree in computer science or similar, and certainly having a degree will open more career doors for you. However, there are a fair number of companies that will hire talented coders even if they do not have a college degree —they are more interested in skill and talent for coding than they are the actual degree. For example, Amazon hires software engineers who do not have a college degree if the person is good at coding and has enough experience to demonstrate that they would be good in the position.

Similarly, anyone can develop an app, so you do not necessarily need a college education to do so. Teenagers have developed apps long before even really thinking about college. For example, Nick D'Aloisio was 16 years old when he came up with the idea for an app that summarizes news content from major news sites to make skimming the news easier to do on smartphones. At the age of 17, he sold his app to Yahoo! and joined Yahoo! as a product manager. If software engineering appeals to your talents and your creative side, you do not necessarily need to wait until you earn a degree to start working on it.

Career Insight:
Don't Stop at Software!

Software engineering is a good way to use your creativity to build a career you love but do not discount hardware engineering, either. Hardware refers to the machines and physical components of a computer or other electronic devices—not the programs that run on them. So, for example, a smartphone is a piece of hardware; an app is software that runs on a smartphone.

Hardware engineering may not sound all that creative but it can be. Virtual reality (VR) is one of the most exciting technologies to come around in the past decade, and one of its great success stories is Palmer Luckey, who designed the Oculus Rift VR headset. He eventually sold Oculus VR, the company he founded, to Facebook (now known as Meta) for $2.3 billion. Luckey was just 17 years old when he built the first prototype for the Oculus Rift in his garage, and 21 when he sold the multibillion-dollar company.

Developing VR technology is a perfect way to combine your creative and engineering talents.

A DAY IN THE LIFE OF
An Audio Engineer

Working with musicians often means keeping long, erratic hours, but the thrill of creating a song makes it all worthwhile. Here is what an average day might look like.

1:00 p.m. You didn't get home until 6:00 a.m. because the band was in the recording studio all night, so you went straight to bed and are just starting your day now.

2:00 p.m. You work on the mix from last night's recording session and it sounds pretty good—you just need to balance some levels and then it'll be great.

3:30 p.m. You check your phone for messages and answer any emails. Once you've done your admin, you'll have the afternoon to yourself, because you're not due back at the studio until much later. Tired, you take a nap.

7:00 p.m. To prepare for the night in the studio, you listen to the tracks again and make some notes so that you can work on some issues with the band.

9:00 p.m. You arrive back at the studio for another night of recording and mixing.

9:30 p.m. As an audio engineer, you know you must place the focus on both the music and technical aspects of the audio session. You adjust equipment in the studio to make sure all the instruments that are being used sound

Being an audio engineer is not a 9-5 job. It involves long nights, and a lot of coffee!

great as well as making sure the singer's voice is being heard and represented authentically. This is a difficult job—you have to balance the art of making a track alongside the science of recording. It takes skill and experience to do that well, but you always enjoy the challenge.

10:00 p.m. You record the first part of the track but you are not convinced it is working. As an audio engineer, you can change the tone, diction, and rhythm of a voice. Having so much control over a singer's voice can sometimes be an issue when working with recording artists. You are aware of this, and always try to build a strong relationship with the artists you work with before any recording begins. You know that they need to feel comfortable with you, and trust that you want to produce the best result you can. Audio engineers need to be confident personalities—you must be able to make your opinions heard. However, you also need to be sensitive to the people that you work with, because music is such an emotional area of work and creativity.

10:30 p.m. After a quick break, you start over. The band lays down two good tracks, one of them requiring only 5 takes but the other requiring 15. You feel like you could mix the tracks in your sleep at this point, but you are hoping to get at least two more done before your studio time ends at 5:00 a.m.

12:30 a.m. The band is starting to get cranky, so you call a short break and order some food.

1:30 a.m. The food made a big difference, and the band members are once again doing their thing, so you call to start the next track.

5:00 a.m. It's a good night—you get two more good tracks laid down and the preliminary mixes started. You leave the studio feeling that you've accomplished something, and head home for some much-needed sleep.

One of the most exciting parts of your job is working at live music festivals.

Creator

Now that you've learned about some of the creator roles in engineering, explore more about how you feel about each job. Use the questions below to help you assess how strongly you feel each role might suit you.

Industrial Designer

- Would you enjoy carrying through your creative vision for a project?
- Would you be happy to balance your creative skills with your technical knowledge?
- How do you feel about working with clients, meeting their briefs, and changing and adapting your work to fit their requirements?
- What would you find rewarding about the job? What might the challenges be? How would you deal with those?
- Would you be happy to collaborate with other designers on projects? Why?

Audio Engineer

- Why are you passionate about music?
- What skills do you have to make you well suited to this job?
- What would you find challenging about working in this job? How would you deal with these challenges?
- Audio engineers often work for themselves. Why would this suit you?
- Do you think you would enjoy working erratic hours and spending your nights working in studios? In what way would this impact your lifestyle?

Software Engineer

- Software engineers can work in any field. What field interests you most and why would you like to work in it?
- This is often a solitary job, why would that suit your personality?
- What strengths could you bring to the job?

Acoustical Engineer

- What skills do you need to do this job? How could you improve on the skills that you aren't strong in?
- Acoustical engineers are often self-employed. What do you think the benefits of that might be? What might the disadvantages be?
- Would you be able to work with a team of architects, engineers, and contractors to ensure that a project runs smoothly and according to its schedule?

Research More

If you like the idea of working as a creator in engineering, but are not sure the jobs covered in this chapter suit you, here are some more creator roles you could explore.

Naval Architect
Technical Coordinator
Spacecraft Designer

CHAPTER FIVE

ORGANIZER ROLES IN ENGINEERING

Engineering is a great field for people who enjoy working in an orderly manner. Engineering also requires much attention to detail. If organization is something that appeals to you, you might want to consider a senior-level or management position in engineering. While those positions do still have a hands-on component, there is a lot of project management that goes on at the higher-level positions, which is perfect for people who thrive on organization. Let's examine some organizer roles.

Engineering Manager

Most industries are organized into two distinct groups: lower-level employees and management. Some people are happy working in lower-level positions within an organization their entire life—they can earn a good living, do work that they enjoy, and not have to worry about managing other people. For some, the idea of managing other employees is daunting. Not everyone is cut out for management. However, others aspire to eventually be a part of an organization's management.

Engineering managers are responsible for overseeing large teams of people and the work that they carry out. They must check that the work is being carried out on time, within budget, and to plan.

They enjoy the challenge of managing the work of multiple employees and building a strong, reliable team that can deliver results.

This breakdown in employee structure is found in all fields of engineering. The engineers spend their days working on their projects, in whatever capacity they have been assigned. The managers also do some of that work but a significant portion of their time is spent looking at the big picture of the project: Are the employees below them finishing their tasks on time? Is the project staying within budget? Are there any roadblocks that may slow down the project or impact the budget? If these big-picture management tasks appeal to your organizational side, then engineering management might be a perfect fit.

Technically, you only need a bachelor's degree and experience in engineering to become an engineering manager. However, in a competitive field, if someone has an advanced degree applicable to engineering management, they are more likely to get the job over someone who has only a bachelor's degree. Applicable degrees for this field include master's degrees in engineering management (MEM) or technology management (MSTM), or a master's in business administration (MBA). These advanced degree programs usually include coursework in human resource (HR) management, accounting and financial management, and engineering economics. Sometimes, depending on the degree program, they may also include courses in project management, quality control, general management principles, and health and safety issues.

While working in a lower-level engineering position, you may also want to position yourself for a move to management, if that is your end goal. You can do so by leading engineering teams at a lower level—in other words, you lead a small team of engineers and report to the engineering manager.

Career Insight:
Entertainment Engineer

There are management roles in any type of engineering, and entertainment engineering is no exception. Entertainment engineers create the structures and the mechanics that are used in theatrical and other performances such as acrobatic performances like Cirque du Soleil, "flying" trapeze effects done on stage by performers such as Pink, and any of the other unbelievable effects done on stage in Broadway shows, the Super Bowl half-time show, and concerts and other live performances. All of these performances require teams of skilled engineers to ensure the safety of participants and that the effects are completed as planned.

Those teams of skilled engineers are managed by people such as Bill Gorlin, who is a vice president for the entertainment division of a major engineering company. Gorlin carefully hires engineers whom he thinks are well suited to the unique challenges of large-scale entertainment engineering, especially mechanical engineers. He is interested in applicants' problem-solving skills and how much they have worked with their hands. As with many engineering roles, Gorlin believes that having a PE licence is a valuable asset.

The engineers involved with Cirque du Soleil help put on an amazing show.

Computer and Information Systems Manager

If your interests lie more in the IT sector than the typical engineering branches of civil and structural engineering, you might want to consider a career as a computer and information systems manager, which is also sometimes referred to as an IT manager.

Similar to engineering managers, IT managers plan and direct the activities of an organization—only in their case, it is the computer-related activities instead of things such as construction projects. A great many companies have IT managers—not just software and other computer-related companies. Any company of a reasonable size that uses computers on a daily basis will probably have an IT manager. For example, a publishing house will use computers for things such as editing manuscripts, creating graphics and page designs for their publications, and tracking information about products, authors, editors, and other contributors in a database. All of these functions require computers and therefore, the publishing house needs someone on its staff who can keep the computer systems running efficiently, upgrade them where processes can be improved, and repair them when problems occur.

Working Your Way Up

In larger companies focused on computer-related activities, there will likely be multiple IT managers, organizing IT teams for different departments. The IT managers would then report into a single vice president (VP) of information technology, chief information officer, or a similar member of management.

IT managers need to have only a bachelor's degree in computer science or a related field, plus significant work experience. However, these positions are quite competitive, so many companies require a master's degree in business or a master's in computer science. And even then, without the relevant experience, you'll have to work your way up to the position.

It can take many years to work up to the highest IT management positions but steady dedication to a particular sector of the IT field will facilitate this.

It's worth noting that taking lower-level IT management positions in a variety of companies is not always helpful for moving up the corporate ladder and becoming an IT manager, unless the companies are all in the same field.

Career Insight:
Women in Engineering

Traditionally, engineering has been a very male-dominated field. A recent survey by the Bureau of Labour Statistics (BLS) showed that only 14 percent of engineers were women and 24 percent of computer scientists were female. However, there has been a real push to try to encourage more women to pursue STEM careers and in particular, engineering careers. The number of women pursuing engineering, math, statistics, and computer science degrees in college more than doubled between 2006 and 2014, according to the National Science Board (NSB) and these numbers look set to continue to rise.

According to the American Society for Engineering Education (ASEE), the top five engineering degrees recently chosen by women were environmental engineering, biomedical engineering, biological engineering, chemical engineering, and architectural engineering. The number of women choosing to study computer science is relatively lower so there are programs in place to encourage women to enter these fields. For example, GirlCode runs the GirlCoder Club, which is a network of volunteer-led coding clubs designed for primary and high school girls who want to have a strong foundation in basic programming skills.

There are also many grants, scholarships, and incentives offered to women who are interested in pursuing engineering degrees, so if you are a woman interested in engineering, do not let the numbers stop you. See what is out there, and go for it!

Talented young female engineers are bringing their skills to the world of engineering and opening up a career pathway for the next generation.

A DAY IN THE LIFE OF

An Engineering Manager

If you have strong organizational skills and are someone who thrives on a challenge, then maybe you can see yourself as an engineering manager. Perhaps this typical day will inspire you.

8:00 a.m. You arrive at the office and after checking your calendar, you remember that you have an early-morning meeting. You've already prepared for it but you go through your notes, just to check everything is in order. When that's done, you spend a bit of time answering emails.

9:00 a.m. You attend the meeting of engineering leads. Your division has 50 engineers, and rather than managing them all, you have assigned an engineering lead to manage each group of five lower-level engineers. Twice weekly, you get together with the leads to find out how their teams are doing.

10:30 a.m. You return to your desk to find 27 emails waiting to be answered, so you set to work answering those before tackling a problem brought up at the meeting with the engineering leads.

11:00 a.m. With all your emails answered, you start addressing the problem: The lead's team is running behind schedule, which is going to put the rest of the project behind schedule. You study the work done by his team to determine why the team's fallen behind, and realize that one of their tasks requires more manpower than five people can handle.

Dealing with team problems requires a calm attitude. You need your team to see that you are in control.

11:30 a.m. You call another engineering lead into your office for a meeting. You explain that one team is running behind and needs more manpower. You ask if you can borrow two members from her team, which is running ahead of schedule, to help the other team catch up. She tells you that she can't spare both team members, but she can spare one. One is better than none...

12:00 p.m. Back to the drawing board. You take the one engineer the team lead can spare and look to the other teams to see whether there's anyone else you can move around to bring the team back up to speed. There isn't, so you set up an appointment to talk to the VP of engineering about possibly bringing in another team member from outside. He's not available to meet until 2:00 p.m., so in the meantime, you get back to your email, which has piled up.

2:00 p.m. The VP of engineering says he's sorry, but there's no budget left in the project to hire another outside engineer. And what's more, the project cannot run late, or you'll face hefty fines. You'll need to find resources within the company to bring the team back up to speed.

3:00 p.m. You pick the team that seems like it will be the least impacted and call the engineering lead into your office to deliver the news. He complains that the rest of his team will have to work overtime to pick up the slack. You apologize but explain that it's necessary —and assure the irritated lead that you, too, will be working over the weekend to help the project catch up. When a project runs behind, it's all hands on deck—even the management. In fact, your work on the project will begin as soon as the lead exits your office.

7:30 p.m. You finally head home, you're tired after a long day. The weekend will be equally long, but the project can't wait. You are feeling positive, though, you know your problem-solving organizational skills are up to the challenge.

Organizer

Now that you've learned about some of the organizer roles in engineering, explore more about how you feel about each job. Use the questions below to help you assess how strongly you feel each role might suit you.

Engineering Manager

- What is it that you find interesting about the role? What interests you least?
- Would you be happy to manage a large team of people and make sure they are completing their jobs well? How would you deal with those who are not?
- Would you be happy to have the responsibility of managing large budgets? Why do you say so?

Computer and Information Systems Manager

- As a junior IT manager, you may spend a large part of your day helping others who aren't very good with computers. What skills would you need to do this well?
- Senior IT managers manage teams. Would this suit you? Why or why not?
- What challenges do you think you might face in this job? Do you have particular skills that might help you deal with them?

"Always keep an eye on the future and how your career might change. For example, will technological advancements change some aspects of your job?"

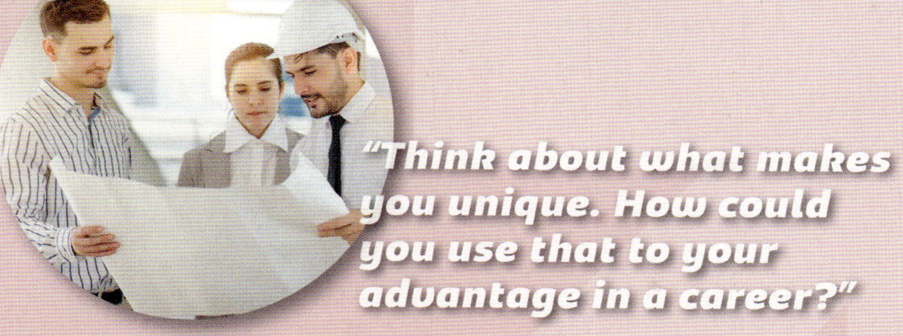

An Engineering Manager

- Review the Day in the Life feature. Did anything about the day surprise you? What was that?
- What do you think the challenges were, and how might you deal with them?
- What aspects of the role do you think you would enjoy?
- Look at the structure of the day and the working hours. Would you be happy with that structure? How might it affect your lifestyle?

Research More

If you like the idea of working as an organizer in engineering, but are not sure the jobs covered in this chapter suit you, here are some more organizer roles you could explore.

Engineering Project Director
Transportation Planner
Assembly Manager

CHAPTER SIX

THINKER ROLES IN ENGINEERING

Thinkers are the people who like to consider every possible angle and really analyze the process down to the finest level of detail. There are a number of engineering positions well suited to the thinkers of this world.

Proof of Concept Engineer

For any new product, someone comes up with the initial idea and brings it to the table. Usually, a development team will discuss the idea and refine it through the process of general brainstorming. But then it comes time to really putting the idea to the test, and that is where proof of concept engineers come into play.

Proof of concept engineers and specialists think about whether that great idea is really practical and feasible. Is it for something that will actually appeal to a market, for example, tiaras for cats might seem like a fun idea but how many people would really buy tiaras for their cats?

Proof of concept specialists always determine what resources are needed to bring the idea to fruition. If the idea is for a product, proof of concept engineers and specialists will determine what materials will be needed to manufacture it, where that manufacturing can be done, and how

To be a proof of concept engineer, you need to be able to think through concepts and figure out if what seems like a great idea can actually become a practical reality.

much it will cost. If the idea is for a type of software, they determine whether the software concept can actually be a reality. For example, VR was not a reality just a few decades ago. It might have been an idea in someone's head but the technology did not exist to make it a reality. Now, that technology does exist, and a proof of concept engineer likely determined how to make the idea become a reality.

The proof of concept is necessary for a few reasons. There is no point in wasting time, money, and human resources if an idea simply is not feasible. The proof of concept helps with funding and most new ideas need funding in some way.

Many companies have to look for grants, loans, or investors to make an idea a reality. And do you know what grantmakers, banks, and investors want to see before giving you money? Proof that the concept is a good idea and will give them a good return on their investment. When it comes to using your proof of concept to secure funding, you need to be a bit of a salesperson. Do your research on your market, why your idea will benefit that market and create a demand, and how you can make it become a reality. And then sell all of that to the people with money, so that you can actually bring your idea to life.

This quality assurance (QA) engineer is checking the manufacturing equipment in a cosmetics factory.

The Bridge

One tech company in Silicon Valley describes its proof of concept engineers as "the bridge between engineering and sales." This is a good description for the reality of what a proof of concept engineer does.

Proof of concept engineers also work on building prototypes for their idea—whether it's a simple version of a software package or a scale model of a product. The prototype can help prove that the idea can really work. Proof of concept is a fairly new segment of the engineering world, but those who want to work as a proof of concept engineer should have at least a bachelor's degree in the field or a related one, as well as several years of experience.

Quality Assurance Engineer

While the terms "quality assurance" (QA) and "quality control" (QC) are often used interchangeably, they mean slightly different things. QA engineers generally look at the process behind making the product, whereas QC specialists typically look at the products themselves to ensure that quality standards have been met. So, for example, in a plant that manufactures bicycles, the QA engineer might look at all the steps of the manufacturing process and streamline them to produce the bicycles more efficiently and with less wasted time and materials. The QC inspector, on the other hand, might inspect each bike and make sure the gears work, the tires are correctly inflated, and the brakes are functioning as they should be.

Typically, a bachelor's degree in the relevant field is needed to become a QA engineer, along with experience in the field. A master's degree is even better but there are QA engineer jobs available for those with only a bachelor's degree. It really depends on the company. For some, if you have relevant computer science experience, that will be enough to assure them that you are capable of doing QA on their team. For most engineering positions, a college degree will never hurt and will open many more doors. Many QA engineer jobs are for software quality assurance. However, there are such jobs available in other engineering fields, too—you may just have to hunt around a bit to find them.

A QC will examine the end product of a production process, such as a food product, to make sure that it meets required industry standards.

There are many QA jobs out there, but they can be hard to get because being a QA is a good career choice—it pays well, and the job stability is generally good. So how do you get into the field?

First of all, you need to be passionate about the product. If you want to prove you are QA material, you should learn anything and everything about the product. You have to know it inside and out to be able to ensure its quality. Then, if you can break the product, all the better. If there is a security hole in a company's software, for example, and you can identify it and share how you would address it, you will be one very large step ahead of the competition for the job.

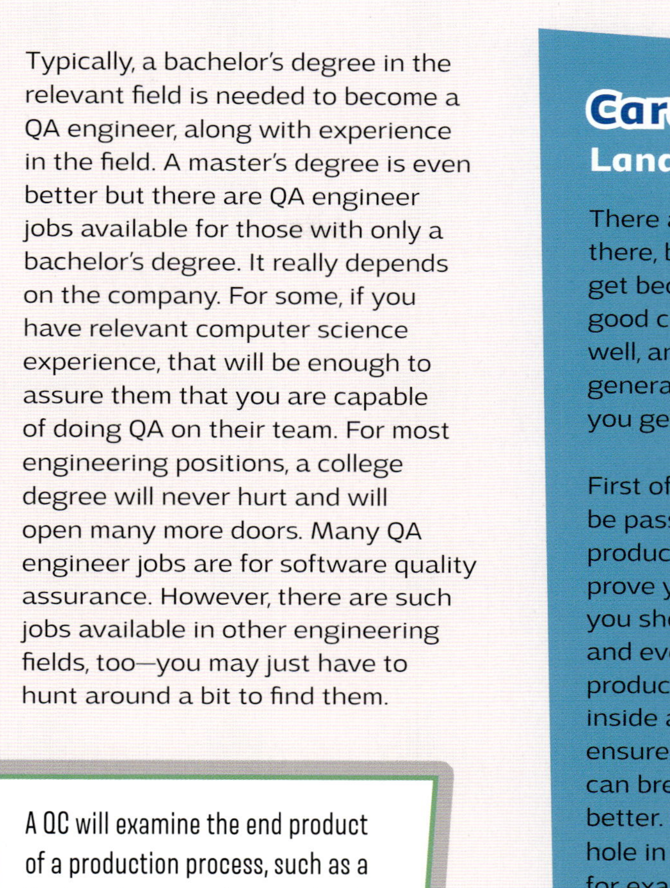

Quality Control Inspector

If earning a degree is not something you want to do right away or you cannot afford to go to college, you might want to consider a QC inspector position. QC inspectors look over products for defects. It is very specific work—you might be looking at the same products all day long, tracking which ones meet specifications and which ones do not—so you need to be organized and methodical. However, this position does not typically require a degree: Usually, a high school diploma and on-the-job training is enough. There are certifications for QC inspectors but usually they are optional, not mandatory. Having such certifications might open up some doors for QC work for you, though.

It's worth noting, though, that one of the difficulties in working as a QC inspector is that the job growth is expected to actually decline in coming years—computers and technology are rendering the job of QC inspectors obsolete in many places. However, if you are interested in QA or QC, this might be a field to look at because it would allow you to gain industry experience and perhaps eventually become a QA engineer.

Does playing video games for a living sound like your dream career?! If you are a super-talented game player, you could turn that much-loved hobby into a career.

Career Insight:
Play Games for a Living?

If you like video games and you want to play them all the time, you might be able to get paid to do this by working as a video game tester. Video game testers are also sometimes called beta testers because they are testing the beta version of the game—that is, the version that comes before the game is released to the general market.

Beta testers play games over and over, looking for bugs and glitches that need to be fixed. It is not the same as just playing for fun, though. Beta testers may have to play the same level many, many times, even long after they have progressed, in order to try different gameplay methods that could produce bugs or glitches. If there are 20 possible ways to slay the dragon, the beta tester has to try them all, just in case there's a problem.

Sometimes, beta testers work in-house at a company but often they just work from home. When they find glitches and bugs, they communicate those to the software development team so they can be fixed.

A DAY IN THE LIFE OF

A Proof of Concept Engineer

Working as a proof of concept engineer is challenging and rewarding. These pages show how a typical day often pans out.

8:00 a.m. You arrive at the office and before you can even open your laptop, you are called into a meeting. You are intrigued.

8:30 a.m. The meeting is about an exciting new idea the design team has had. It's a robot to feed people's pets. You listen to the team haggle over the pros and cons of such a product. Pro: no more pet sitters or expensive kennels needed when people travel—the robot can feed their pets. Con: the robot is electronic, so it probably can't give the pets fresh water—water is, after all, the death of electronics! So a pet sitter would still be needed. Also, could the robot let the animal out when it needs to go out? Again, a pet sitter would still be needed. Pro: the robot could also feed the animals when their owners are home, so the animals wouldn't wake their owners at 5 a.m. for food. Con: while that's a plus, would owners really want to pay for such a product? Or would they rather just get up and feed their pets themselves? And the debate goes on and on. Ultimately, the team decides it would like to see a proof of concept.

You love investigating robotics—it's your favorite research area.

10:30 a.m. You return to your desk to find 12 emails waiting to be answered, so you set to work answering those, all the while you're thinking about the robotic pet sitter.

11:30 a.m. You take an early lunch break, because the afternoon will be spent doing market research.

12:30 p.m. You start researching the market for the pet-feeding robot. It is your job to ask a lot of questions: Are similar products available already? How are they selling? How much are they selling for? What are the customer reviews on them? What features do they lack that your team could improve on with their robot pet feeder?

2:30 p.m. With your research gathered, you begin writing the market analysis section of the proof of concept report. It's a time-consuming process because the report is not just a proof of concept, but will also be used as a sales tool if the robot feeder gets to that stage of development.

5:00 p.m. You jot down ideas for the next section of the report, on the materials that could be used for such a product, and how much they might cost. You haven't had enough time to research it fully yet, but you want to get your ideas down before you leave the office for the day. Tomorrow will begin another full day of research. You still need to research what sort of computer hardware will be needed to run the robot, as well as what software might best work. You'll also need to map out the potential problems that could come up with the product and how to address them. For example, there's the problem of making the robot waterproof, because where there's pet food, there's often a dish of water for the pet. The robot is bound to get wet even if you instruct users not to use it for giving pets water! And then of course you have to get started on a prototype. It's going to be a long several weeks of work on this robot pet feeder, but you've got a good feeling that it will be worth your effort.

3:30 p.m. You begin researching the next phase of the report: what materials would be best suited for such a product? This requires looking at other robot products on the market and seeing what they use— there is no point in reinventing the wheel when certain materials are already being successfully used in robotic products.

Building prototypes helps you work through the possible problems of any products.

Thinker

Review and Check In

Now that you've learned about some of the thinker roles in engineering, explore more about how you feel about each job. Use the questions below to help you assess how strongly you feel each role might suit you.

Proof of Concept Engineer

- In this job, you'll need to communicate effectively with developers, management, and other workers. What communication skills do you have? How can you improve on them?
- What would you find rewarding about the job?
- What might the challenges be? How would you deal with those?

Quality Assurance Engineer

- This can be a hard job to land. What are you willing to do to make sure you get the job you apply for?
- You have to be analytical and decisive to do this job. What other strengths do you think you could bring to this role?

"*Close your eyes and picture yourself doing your ideal job in the future. What type of workplace setting are you in and what are you doing? Keep that vision in mind as you research careers.*"

Quality Control Inspector

- What aspects of the job do you think you would enjoy? Why?
- Are there parts of this role that you would find challenging? Would you enjoy them?
- Would you be happy to work on the same product for a long period of time or do you think you would find this monotonous?
- How do you feel about the decline in job growth in the field? How could it affect you in the future?

Research More

If you like the idea of working as a thinker in engineering, but are not sure the jobs covered in this chapter suit you, here are some more thinker roles you could explore.

Aviation Accidents Investigator
Composites Engineer
Production and Process Engineer

If you have come to the end of this book and are ready to start making a career in engineering a reality, follow the checklist on these pages to kick-start your future.

All roles in engineering will require strong math and science skills, so focus on your classes at school, paying particular attention to technology and engineering. If your school offers computer science and IT courses as options, make sure you sign up for them. In those courses, you'll become familiar with many different aspects of the type of skills that could be useful in engineering, such as computer-aided drafting and design (CADD) and robotics. And get involved in any engineering clubs at your school or outside of it—gaining experience now will set you up for a future career.

Creator roles in engineering will allow you to mix your creative talents with your engineering skills. Studying visual arts and computer science at school will help you focus on these two areas.

Talk to a Guidance Counselor

If your guidance counselor offers career advice, take it! Guidance counselors will be able to help you explore a lot of different options and talk more about whether a role in engineering could suit you. They will also be able to advise you on further education courses you could pursue after school or in-job training that could suit you.

Your parents and parents' friends may have great contacts in the engineering world. They may be able to put you in touch with an employer, so that you can talk to them directly about roles in engineering that interest you. For example, a parent or friend could connect you with a mechanical engineering company, so that you can talk to the people who work there about the roles they carry out.

Walk in Their Shoes

Some employers are happy to have students meet with them in person for informational interviews, through which you can find out what people in engineering roles do. You may even be able to shadow someone who works in engineering. For example, you could spend a day or two at work with an industrial designer, finding out more about what they do and how they do it. That's a really great way to discover if you would like to do their job in the future.

Make Your Summer Count

Getting a summer job is a great way to get experience, even if it is at the very bottom of the ladder. Working in any engineering-related company, no matter how basic your role is to start with, will give you valuable experience that will help you in the future. For example, just through helping out in a car repair shop that fixes and services cars or bicycles, you will discover more about engineering and if a career in this area might suit you.

Get Experience

The best way to find out if a career is really going to suit you is to try it out! Many businesses offer internships, which are unpaid temporary jobs in which people can get work experience. Internships usually run over the school summer break, to give young people the best chance of trying out a role over a longer period of time. Most employers only take on interns who are 16 years or older, but sometimes internships have been offered to students as young as 14 years old. An internship will look great on your resume when you come to apply for jobs in the future, and some internships even come with an academic credit if they are built into an educational program. Ask your guidance counselor if such an educational program is available at your school.

Do Your Research

You can never do enough research when it comes to planning a future career, so explore as many resources as you can to find out more. Start by taking a look at the resources on page 63 of this book to learn more. For example, the U.S. BLS offers some great resources that can help you learn more about possible roles in engineering. Check out in particular their Occupational Outlook Handbook section.

As you research, remember there is never a right or wrong in making choices, and you can always change your mind. Keep flexible, be positive about your future, and have fun choosing your perfect STEM career.

GLOSSARY

acoustical related to sound

adrenaline a hormone that is secreted by the adrenal glands in times of stress

analysis detailed examination of something, usually for the purpose of drawing conclusions

apprenticeship a position in which a person works for lower wages while learning a trade from a skilled professional

associate's degree a two-year degree awarded by a community college

biomass organic matter that can be used to generate power

cholera an infectious bacterial disease of the intestine

dysentery an infectious disease of the intestine that can be fatal

erratic irregular or unordered

facilitate to make something easier

feedback a screeching sound made when an amplifier or microphone is too close to its input

fossil fuels natural fuels formed from the remains of living organisms

geothermal energy energy produced by Earth's internal heat

hardware the physical parts of a computer, including machines and wiring

hydropower power obtained by harnessing the energy of water

implementing putting into place

incentives things that motivate people to do something

methodical orderly behavior

mixing board an electronic device that allows a sound engineer to mix sounds from different sources

nonprofit an organization in which virtually all profits are invested back into the organization, usually to provide services that benefit people

obsolete out of date and no longer needed

optimal the best

PhD a doctoral degree in a field other than medicine

prosthetics artificial body parts, such as limbs or certain organs

prototypes initial versions of something, usually for testing

refugee a person who must leave their home country to escape war, persecution, or a natural disaster

renewable a source of power that is not depleted by use

software the programs and operating system used by a computer

solar cells devices that convert energy from the sun into electricity

solar panels devices that absorb the energy of the sun

solar power power obtained by harnessing the energy of the sun

sustainable able to be maintained without causing damage to the environment

therapeutic related to healing

virtual reality (VR) a computer-generated simulation of an environment that users can interact with

vocational schools schools that are designed to teach the skills necessary for certain occupations

wind power power obtained by harnessing the energy of the wind

Books

Currie-McGhee, Leanne K. *A Career in Mechanical Engineering* (Careers in Engineering). Referencepoint Press, 2018.

Dahnke, Sarah Rose. *Careers for Tech Girls in Software Engineering* (Tech Girls). Rosen Young Adult, 2018.

Kallen, Stuart A. *A Career in Environmental Engineering.* (Careers in Engineering). Referencepoint Press, 2018.

Websites

Take a look at the BLS site for more careers guidance:
www.bls.gov/k12/students/careers/how-can-bls-help-me-explore-careers.htm

Check out the BLS Occupational Outlook Handbook to find out more about different jobs and the qualifications you need for them:
www.bls.gov/ooh

If a software engineering career is your goal, Code.org is a great place to start learning how to code. Log on for free at:
https://code.org

If environmental engineering is your interest, this site for teachers and students, produced by the United States Environmental Protection Agency (EPA), is full of great resources and activities:
www.epa.gov/students

If your interests lie in aerospace engineering, you'll love NASA's website for students:
www.nasa.gov/audience/forstudents

This website connects teens with experiential learning opportunities, including summer programs, community service opportunities, and other programs:
www.teenlife.com

Publisher's note to educators and parents:
All the websites featured above have been carefully reviewed to ensure that they are suitable for students. However, many websites change often, and we cannot guarantee that a site's future contents will continue to meet our high standards of educational value. Please be advised that students should be closely monitored whenever they access the Internet.

acoustical engineers 32, 39
aerospace engineers 25, 29
Amazon 34
apprenticeships 23
architectural engineering 45
audio engineers 31, 36–37, 38
automotive service technicians 25
biomedical engineers 20, 28, 45
budgets 40, 41, 47, 48, 51
builder personality and roles 5, 6, 8, 20–21, 22, 23, 24, 25, 26, 27
certification 24, 32, 54
chemical engineering 45
Cirque du Soleil 42
civil engineering technicians 21
civil engineers 21, 22, 26–27, 28
coding 33, 34, 45
college 4, 8, 14, 23, 25, 30, 31, 33, 34, 45, 53, 54
community college and classes 15, 23, 30
computer and information systems managers 43, 44, 48
computer science at school 60
considering challenges of the job 18, 19, 28, 38, 39, 48, 49, 58, 59
considering rewards of the job 18, 19, 38, 58
considering working as part of a team 28, 38, 39, 48
considering your finances 15, 18, 21, 39
considering your lifestyle 15, 19, 29, 38, 39, 49
considering your skills 18, 19, 28, 29, 39, 48, 58
considering your values 39
considering your work environment 5, 28, 29, 39
creator personality and roles 5, 30–31, 32, 33, 34, 35, 36, 37, 38, 39
degrees 4, 13, 14, 15, 20, 21, 23, 25, 30, 32, 33, 34, 41, 44, 45, 52, 53, 54
developing countries 11, 12
diplomas 23, 54
engineer educators 14, 18
engineering clubs at school 14, 45, 60
engineering in medicine 10, 11, 20, 28, 45
engineering managers 40, 46–47, 48, 49
engineering problem solvers 10
English classes at school 60
entertainment engineers 42
environment engineers 13, 18, 45
exploring options 9, 55
flowcharts 5, 6–7, 59
GirlCode 45
grants and scholarships 45
guidance counselors 60, 61
hardware engineering 35
helper personality and roles 5, 6, 10–11, 12, 13, 14, 15, 16, 17, 18, 19
high school 14, 15, 23, 31, 45, 54, 60
human resources (HR) 41
industrial designers 30, 38, 61
informational interviews 61
internships 61
IT courses at school 60
licenses 21
making connections 60
math classes at school 60
mechanical engineers 23, 24, 29, 51
on-the-job training 23, 25, 54, 60
online tuition 31
organizer personality and roles 5, 7, 40–41, 42, 43, 44, 45, 46, 47, 48, 49, 50, 51, 52, 53
personality types 4, 5, 6–7, 8, 9
proof of concept engineers 50, 51, 52, 56–57, 58
quality assurance engineers 52, 54, 58
quality control inspectors 54, 59
renewable energy engineering 13
research 9, 19, 29, 39, 49, 59, 61
resumes 61
robotics engineering 14, 24, 56, 57, 60
salaries 15, 18, 21, 40, 53
schedules 17, 40, 41, 46, 47
science classes at school 60
self-employment 39
software engineers 33, 34, 35, 39
structural engineers 16–17, 19
summer engineering camps 15
summer jobs 61
switching careers 9
technical school 25
the environment 10, 13, 14, 15, 18, 27, 45
thinker personality and roles 5, 7, 50–51, 54, 55, 56, 57
video game testers 55
virtual reality (VR) 35, 51
visual arts at school 60
vocational courses at school 60
vocational school 21, 23, 25
women in engineering 45
working hours 16–17, 26–27, 36–37, 38, 39, 46–47, 49, 56–57
Yahoo! 34
young engineers 34, 35, 45

About the Author

Cathleen Small has written many books for young people on a wide variety of topics. In writing this book she has learned the value of personality testing, research, and considering many options when exploring a future career.